PERENNIAL MATH

Tournament Tests

Grades 7-8

PERENNIAL MATH TOURNAMENT TESTS

These are a collection of tests from past Perennial Math Tournaments. Each tournament has two rounds of testing. Each student takes an Individual Test during the first round and then students combine with their teammates to take a Team Test during the second round. The solutions included in this book are only a short sample of the many possible solutions that could be used to solve each problem.

INDIVIDUAL TESTS GRADES 7-8

30 minute practice tests for individual students

Grade 7-8 Individual Event (30 Minutes)

1) How many rectangles can be made with an area of 144cm whose sides are whole numbers?

2) What number should X be?

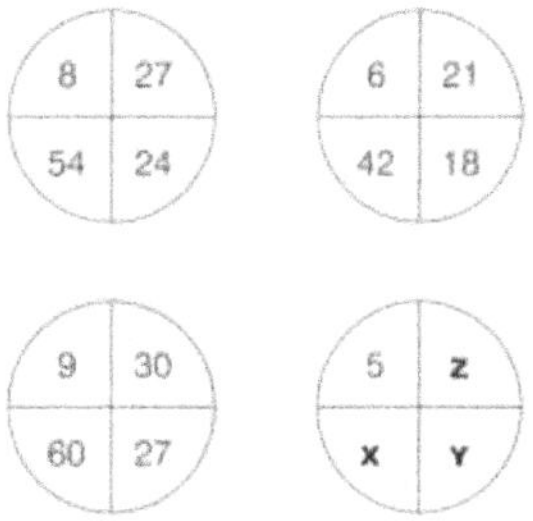

3) How many triangles of any size are in this shape?

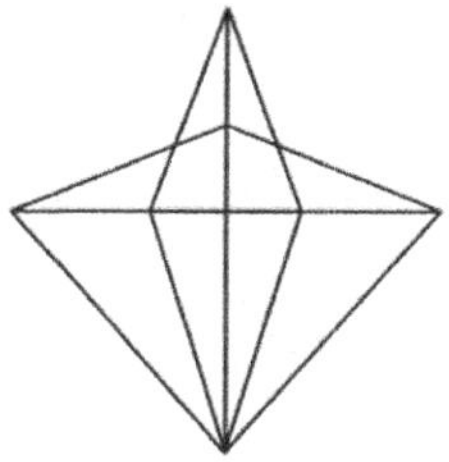

4) It takes you 5 minutes to cut out 7 stars. How long would it take to cut out 630 stars? Reduce answer to hours in decimal form.

5) How many times does the digit 2 appear between 100 and 200 inclusive?

6) Vickie and Amy made $23.10 together selling cookies. Vickie made twice as much as Amy. How much did Vickie make?

7) In this addition problem, different letters represent different digits. If A is less than B, how many solutions are there for this problem?

```
 ABA
+BAB
 CCC
```

8) The auditorium has 2000 seats with 2/5 of the seats reserved for season ticket holders. There are 400 season ticket holders and some non-season ticket holders seated making the auditorium only 3/4 full. How many non-season ticket holders are in the auditorium right now?

9) Of the cookies in the bakery, 1/3 are sugar, 60 are chocolate chip, and the remaining 40% are cinnamon. How many cookies are in the bakery?

10) It is 7 a.m. now. What time will it be in 357 hours? Specify a.m. or p.m. in your answer.

11) 4.0×10^6

12) $[(4 + 12 \div 4) - 2]^3$

13) Find the slope of the line segment joining the points (3, 7) and (- 3, 2).

14) $8 + 2 \times 5 + 4 \times 3 =$

15) What is 40% of 25% of 400?

INDIVIDUAL ANSWER DOCUMENT
(Grades 7-8)

SCORE

Team # __________ School ______________________________

Student Name ____

Student Number ________________

1. ____________	6. ____________	11. ____________
2. ____________	7. ____________	12. ____________
3. ____________	8. ____________	13. ____________
4. ____________	9. ____________	14. ____________
5. ____________	10. ____________	15. ____________

PERENNIAL MATH

Grade 7-8 Individual Event Solutions

1) The area of a rectangle = length x width. List out the factors of 144. (1, 144), (2, 72), (3, 48), (4, 36), (6, 24), (8, 18), (9, 16), (12, 12). So there are 8 possible rectangles.

2) The numbers from top left to bottom right diagonal are a fraction equal to 1/3. Then the number in the top right equals the bottom right plus 3 - then the diagonal from top right to bottom left is a fraction equal to 1/2.

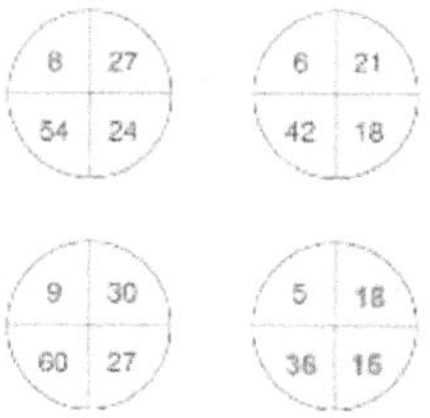

3) There are 22 triangles in the shape. 8 individual triangles made of only one shape; 7 triangles made from 2 shapes combined; 2 triangles made from 3 shapes combined; and then 5 triangles made from 4 shapes combined.

4) Use ratios. 5:7 = n:630
Cross multiply to get n = 450 minutes. Reduced to 7.5 hours.

5) The digit 2 appears 21 times between 100-200 inclusive. 102, 112, 120, 121, 122, 123, 124, 125, 126, 127, 128, 129, 132, 142, 152, 162, 172, 182, 192, 200. Don't forget to count it twice in the number 122.

6) Use algebra. Let A represent Amy and Vickie is equal to 2A. 2A + A = \$23.10. 3A = \$23.10. Divide both sides by 3 and A = \$7.70. Vickie made twice this much, so she made 2(7.70) = \$15.40.

7) If A = 1, then B can be 2, 3, 4, 5, 6, 7, or 8 which is 7 solutions. If A = 2, then B can be 3, 4, 5, 6, or 7 which is 5 solutions. If A = 3, then B can be 4, 5, or 6 for 3 solutions. If A = 4, then B can only be 5 for 1 solution. There are a total of 16 possible solutions.

PERENNIAL MATH

Grade 7-8 Individual Event Solutions

8) Draw a picture. Divide the auditorium into fifths and then separate these into fourths for 20 subareas. Each fourth holds 100 since 100 x 20 = 2000. The auditorium is 3/4 full which is 15/20. So, there would be 15 subareas filled and 5 empty. There are 4 subareas of season ticket holders and 11 subareas of non-season ticket holders. So, there are 11 x 100 = 1100 non-season ticket holders.

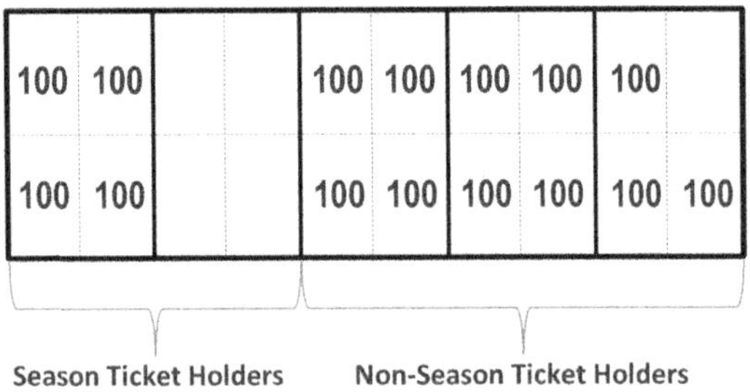

9) There are three groups of cookies. Convert them to like terms to determine the proportion of the 60 chocolate chip cookies. 1/3 + 40% = 1/3 + 4/10 which converted to like terms is 5/15 + 6/15 = 11/15. So the 60 cookies make up 4/15 of the total cookies. Each proportion is 60/4 or 15 cookies. There are 15 proportions of 15 cookies each. 15 x 15 = 225 cookies.

10) Divide 357 hours by 24 hours and there are 14 days with 21 hours remaining. It is 21 hours past 7a.m. The time is 4 a.m.

11) $4.0 \times 10^6 = 4{,}000{,}000$

12) Use the order of operations PEMDAS.

$$[(4 + 12 \div 4) - 2]^3$$
$$[(4 + 3) - 2]^3$$
$$[7 - 2]^3 = [5]^3 = 125$$

13) Find the slope of (3, 7) and (-3, 2). Use the formula: $m = \frac{y2 - y1}{x2 - x1}$

Let $x_1 = 3$, $y_1 = 7$, $x_2 = -3$, and $y_2 = 2$

$$m = \frac{2 - 7}{-3 - 3}$$
$$m = \frac{-5}{-6}$$
$$m = \frac{5}{6}$$

14) Use the order of operations PEMDAS.

$$8 + (2 \times 5) + (4 \times 3) =$$
$$8 + 10 + 12 = 30$$

15) 40% of 25% of 400 = .4 x .25 x 400 = 40

PERENNIAL MATH

INDIVIDUAL ANSWER KEY

(Grades 7-8)

Instructions for Scoring:

- ***Mark CORRECT answers by circling "O" the question number, and INCORRECT answers with an "X". Each correct answer is worth one point. Count the circles (correct answers) and write that number in the "SCORE" box.***

CORRECT → (1.) ________

INCORRECT → X. ________

1. **8**
2. **36**
3. **22**
4. **7.5**
5. **21**
6. **15.40**
7. **16**
8. **1100**
9. **225**
10. **4 a.m.**
11. **4,000,000**
12. **125**
13. **5/6**
14. **30**
15. **40**

PERENNIAL MATH

Grade 7-8 Individual Event (30 Minutes)

1) It costs 17¢ to mail a postcard and 25¢ to mail a letter. If Dykov spent $4.35 and wrote postcards and letters to 19 friends, how many postcards did he mail?

2) The cubes are glued together and dropped in a bucket of paint. How many cubes would have 3 sides painted if the shape came unglued?

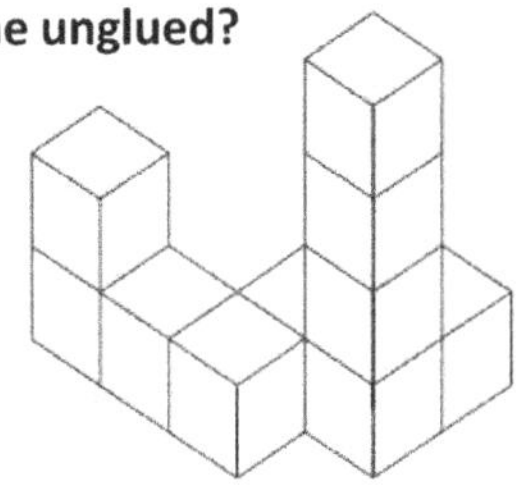

3) How many triangles of any size are in this shape?

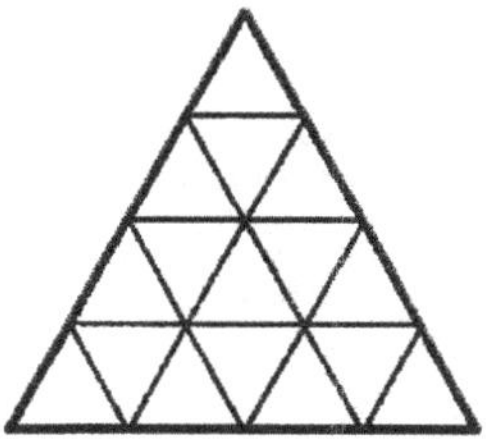

4) If 6 zinks = 2 zarks, and 7 zarks = 1 zelk, how many zinks = 2 zelks?

5) Of all the numbers between 700 and 800 inclusive, how many times does the digit "7" appear?

6) A barrel is half full of oil. Twenty pints are removed and now the barrel is 1/3 full. How many pints will need to be added back in to make the barrel 3/4 full?

7) In this addition problem, different letters represent different digits. What digit does the letter B represent?

$$\begin{array}{r} ABC \\ \underline{+CBA} \\ 564 \end{array}$$

8) Delilah went to three shops. She spent 1/4 of her money in the first shop. She then spent 1/3 of the remaining money in the second shop. When she left the third shop, she had $6 in her purse which was half of the money she had when she entered the third shop. How much money did she have when she first started shopping?

9) A rectangular garden is 25 meters wide by 40 meters long. There is a sidewalk 4 meters wide going around the outside perimeter of the entire garden. What is the area of this sidewalk?

10) What is the sum of the first 10 prime numbers?

PERENNIAL MATH

11) Solve for x. $37x - 106 = 671$

12) $[(4 + 12 \div 4) - 2]^3$

13) Find the slope of the line segment joining the points (1, - 4) and (- 4, 2).

14) Find the volume of a square pyramid of base length 12cm and height 15cm.

15) Find the perimeter of the rectangle whose length is 4 times its width and its area is 144.

INDIVIDUAL ANSWER DOCUMENT
(Grades 7-8)

SCORE

Team # __________ School ______________________________________

Student Name____

Student Number_________________

1. ____________ 6. ____________ 11. ____________

2. ____________ 7. ____________ 12. ____________

3. ____________ 8. ____________ 13. ____________

4. ____________ 9. ____________ 14. ____________

5. ____________ 10. ____________ 15. ____________

PERENNIAL MATH

Grade 7-8 Individual Event Solutions

1) Make a table. Start with 8 postcards and 11 letters. Increase the number of letters until the cost is \$4.35. He mailed 5 postcards.

Postcards (0.17)	8	7	6	5
Letters (0.25)	11	12	13	14
Total Cost	4.11	4.19	4.27	4.35

2) The shape is made of 10 cubes. Look for cubes that have 3 sides glued. There is only 1 cube that has only 3 sides painted.

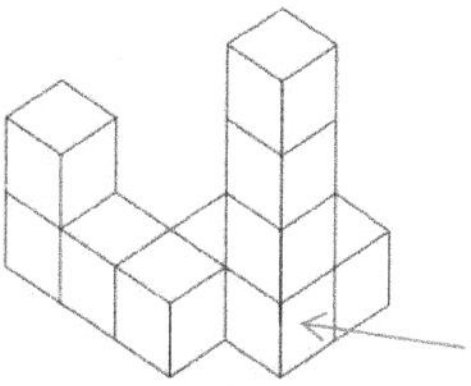

3) There are 26 triangles in the shape. 16 individual smaller triangles. 6 triangles made from 4 triangles combined. 3 triangles made from 9 triangles combined. Then finally 1 large triangle made from all 16 triangles combined.

4) If 6 zinks = 2 zarks, then 3 zinks = 1 zark. Then 7 zarks = 21 zinks = 1 zelk. So there would be 42 zinks in 2 zelks.

5) The digit 7 appears 100 times in the hundreds place between 700-800 inclusive. The digit 7 appears 10 times in the tens place. The digit 7 appears 10 times in the ones place. So the digit 7 appears 120 times between 700-800 inclusive.

6) The twenty pints of oil that were removed was 1/6 of the capacity of the barrel because 1/2 - x = 1/3, or x = 1/2 - 1/3 or 1/6. So 20 x 6 is 120 pints of oil for the entire capacity of the barrel. 1/3 of 120 is 40 pints which is what is in the barrel now. 3/4 of 120 is 90 pints. So 90 - 40 = 50 pints of oil that would have to be added to the barrel to be 3/4 full.

7) Since B + B is even, then C + A could not be greater than 9 or a one would carry from the units place. A and C must be 3 and 1, because they are different numbers and they cannot be 4 or 0. B must be 3 or 8 to yield 6 in the tenths place. Since the hundreds digit is odd, then 1 was carried from the tens place. So B must = 8.

8) Work backwards. If she had \$6 left in the 3rd shop and that was ½ the money, she had \$12 entering the third shop. If she spent 1/3 of her money in the second shop and ended with \$12, then \$12 is 2/3. So 1/3 is equal to \$6. So she had \$18 in the 2nd shop. She spent ¼ of her money in the 1st shop and was left with \$18 so \$18 is ¾ and ¼ would be \$6. So she spent \$6 in the first shop giving her a total of \$24 to start with.

9) Draw the area. The entire area would be 48m x 33m = 1584 m^2. Now subtract the area of the garden (40m x 25m = 1000m^2) to get the area of the sidewalk:

1584 – 1000 = 584m^2

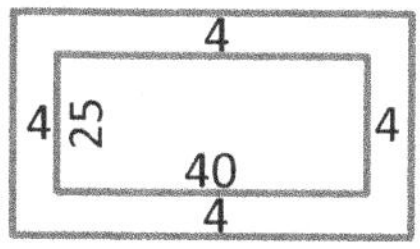

10) 2 + 3 + 5 + 7 + 11 + 13 + 17 + 19 + 23 + 29 = 129

PERENNIAL MATH

Grade 7-8 Individual Event Solutions

11) $37x - 106 = 671$

$37x = 671 + 106$

$37x = 777$

$x = 777/37$

$x = 21$

12) Use the order of operations PEMDAS.

$[(4 + 12 \div 4) - 2]^3$

$[(4 + 3) - 2]^3$

$[7 - 2]^3 = [5]^3 = 125$

13) Find the slope between points (1, - 4) and (- 4, 2). Use the slope formula: $m = \frac{y2 - y1}{x2 - x1}$

Let $x_1 = 1$, $y_1 = -4$, $x_2 = -4$, and $y_2 = 2$.

$m = \frac{2+4}{-4-1}$

$m = \frac{6}{-5}$

$m = -\frac{6}{5}$

14) Volume of a square pyramid is $1/3\ b^2h$

$1/3 * (12cm)^2 * 15cm$

$1/3 * 144cm^2 * 15cm$

$2160/3$

$720cm^3$

15) Since the area is 144, then the length times width = 144. The length is four times the width. So the length is 24 and the width is 6. Therefore the perimeter of the rectangle is 2(24 + 6) = 60.

INDIVIDUAL ANSWER KEY

(Grades 7-8)

Instructions for Scoring:

- ***Mark CORRECT answers by circling "O" the question number, and INCORRECT answers with an "X". Each correct answer is worth one point. Count the circles (correct answers) and write that number in the "SCORE" box.***

CORRECT → (1.) ________

INCORRECT → X. ________

1. 5
2. 1
3. 26
4. 42
5. 120
6. 50
7. 8
8. 24
9. 584
10. 129
11. 21
12. 125
13. - 6/5
14. 720
15. 60

PERENNIAL MATH

Grade 7-8 Individual Event (30 Minutes)

1) Suppose five days before the day after tomorrow was Tuesday. What day was yesterday?

2) What will the 40th number be in column D?

A	B	C	D	E	F
1	2	3	4	5	6
12	11	10	9	8	7
13	14	15	16	17	18
24	23	22	21	20	19
25	26	27	28	29	30

3)The numbers 1 through 7 are placed in each square so that the sums in the column and row are equal. What is the value of A?

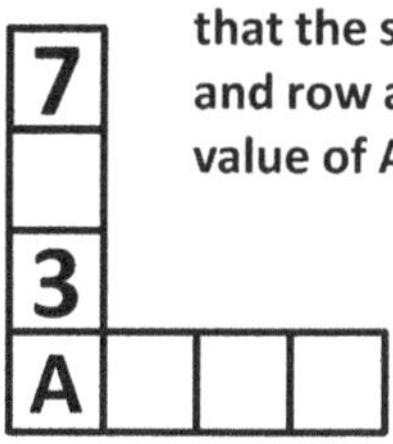

4) Anna has an average of 85 on 4 different tests. If the lowest score is dropped, her average rises to 90. What is her lowest score?

5) Bob has a collection of 33 coins consisting of nickels, dimes, and quarters that has a value of $3.30. If there are three times as many nickels as quarters, and one-half as many dimes as nickels, how many nickels does he have?

6) What is the smallest number that is divisible by 3 and by 5 and leaves a remainder of 3 when the number is divided by 8?

7) Ralph's mother is 3 times as old as Ralph is now. In 5 years, Ralph's mother will be 2.5 times as old as Ralph will be then. What is the youngest age Ralph can be now in years?

8) Craig spent two thirds of his money. He lost two-thirds of the remainder and then had $4 left. With how much money did he start?

9) How many prime numbers less than 100 contain the digit 1 or 3?

10) The product of three consecutive integers is 5,814. What is the sum of the three integers?

11) If x □ y is defined as $(x + y)^2 - 2xy$, then 6 □ 8 =

12) How many different squares are in this shape?

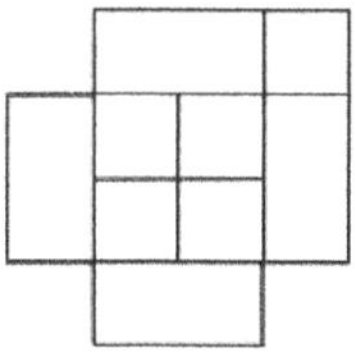

13) The sum of 4 consecutive even integers is 116. The sum of the digits of the smallest of the 4 integers is ________.

14) 5 + 4 x 7 – 8 x 9 =

15) 160 + 150% of 160 =

INDIVIDUAL ANSWER DOCUMENT
(Grades 7-8)

SCORE

Team # __________ School ______________________________

Student Name ____

Student Number _______________

1. ____________ 6. ____________ 11. ____________

2. ____________ 7. ____________ 12. ____________

3. ____________ 8. ____________ 13. ____________

4. ____________ 9. ____________ 14. ____________

5. ____________ 10. ____________ 15. ____________

PERENNIAL MATH

Grade 7-8 Individual Event Solutions

1) Make a number line to help record the days of the week. Yesterday was Thursday.

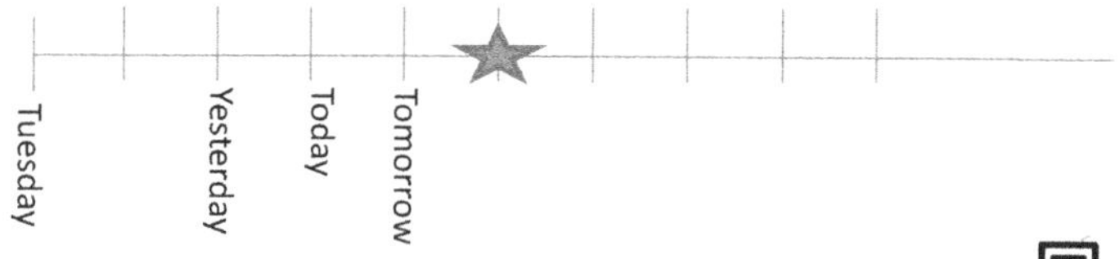

2) The pattern under the "D" column is + 5 + 7, repeated. So if this were carried out to 40 rows, it would be the starting number 4 + 19 pairs of (5 + 7) + an additional 5 in the 40th row. The number under column D would be 237.

A	B	C	D	E	F
1	2	3	4	5	6
12	11	10	9	8	7
13	14	15	16	17	18
24	23	22	21	20	19
25	26	27	28	29	30

...
...
...
237

3) The sum of the numbers 1 through 7 is 28. The sums of the column and row excluding A are equal. If the corner number is 4, then other numbers can be combined to get the same sums:
7 + 3 + 2 = 1 + 6 + 5.
A is equal to 4.
If the corner number is 6, then other numbers can be combined to get the same sums: 7 + 1 + 3 = 2 + 4 + 5
A can also be equal to 6.

4) Anna has an average of 85 on 4 different tests so the total sum of her tests is 340. If her average rises to 90 when the lowest score is dropped, the sum of her tests will be 3 x 90 = 270. Subtract the 3 test sum from the 4 test sum, 340 - 270 = 70. Her lowest score is 70.

5) Use algebra and express all coins in terms of quarters. Let q stand for the number of quarters: *q*; number of nickels: 3*q;* number of dimes: (3/2)*q*
There is a total of 33 coins, so solve for *q*:

$q + 3q + (3/2)q = 33$

$4q + (3/2)q = 33$ (multiple both sides by 2)

$8q + 3q = 66$

$11q = 66$

$q = 6$

Then there are six quarters, and you can work backwards to figure out that there are 9 dimes and 18 nickels.

6) The smallest number that is divisible by 3 and by 5 must be a multiple of 15. So make a list of multiples of 8 adding 3: 11, 19, 27, 35, 43, 51, 59, 67, <u>75</u>. The number 75 is the smallest number that meets the requirements.

Grade 7-8 Individual Event Solutions

7) Make a table. Ralph's mother is 3 times as old as Ralph is now. In 5 years, Ralph's mother will be 2.5 times as old as Ralph will be then. Ralph is 15 years old now.

Ralph Now	Mom Now	Ralph + 5 Yrs	Mom + 5 Yrs.	Mom/Ralph
5	15	10	20	2
6	18	11	23	2.0909091
7	21	12	26	2.1666667
8	24	13	29	2.2307692
9	27	14	32	2.2857143
10	30	15	35	2.3333333
11	33	16	38	2.375
12	36	17	41	2.4117647
13	39	18	44	2.4444444
14	42	19	47	2.4736842
15	45	20	50	**2.5**

8) Solve with a diagram. Break up the total amount into thirds, and the last third into thirds also with one of those thirds representing the $4 left. The other thirds are $4 as well, so one third of the total amount is $12. $12 x 3 = $36 for the starting total.

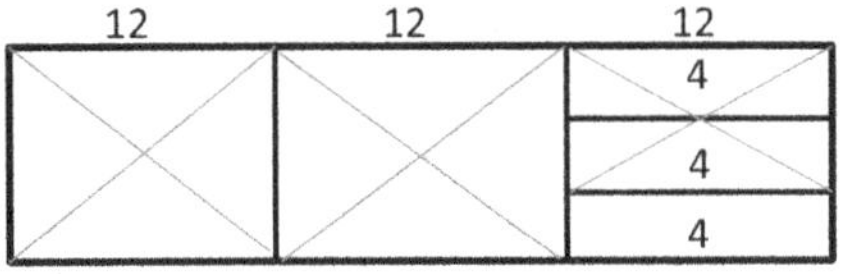

9) The prime numbers that contain the digit 1 or 3 are 3, 11, 13, 17, 19, 23, 31, 37, 41, 43, 53, 61, 71, 73, and 83. There are 15 prime numbers in this set.

10)Use factoring because the question asks for the product which is the opposite of factoring. The factors of 5,814 = 2 x 3 x 3 x (323) and 323 is 17 x 19. This is helpful because we are looking for three consecutive numbers. So the integers are 17, 18 and 19 and their sum is 17 + 18 + 19 = 54.

11) If $x \square y = (x + y)^2 - 2xy$, then $6 \square 8 = (6 + 8)^2 - 2 \times 6 \times 8 = 196 - 96 = 100$.

12) There are 11 squares.

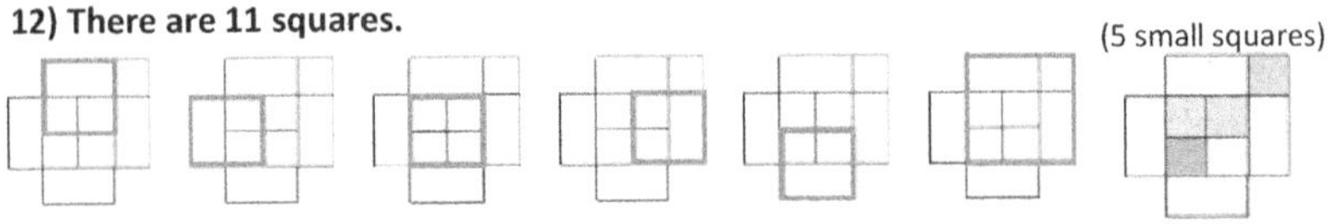

13) The sum of 4 consecutive numbers is 116. The average of the 4 numbers is 116/4 = 29. So the four numbers are 26, 28, 30 and 32. The sum of the digits in the lowest number is 2 + 6 = 8.

14) 5 + 4 x 7 – 8 x 9 =
5 + 28 – 72 = -39

15) 160 + 160 x 1.5 = 160 + 240 = 400

INDIVIDUAL ANSWER KEY

(Grades 7-8)

Instructions for Scoring:

- ***Mark CORRECT answers by circling "O" the question number, and INCORRECT answers with an "X". Each correct answer is worth one point. Count the circles (correct answers) and write that number in the "SCORE" box.***

CORRECT → (1.) ________

INCORRECT → ~~2.~~ ________

1. Thursday
2. 237
3. 4 or 6
4. 70
5. 18
6. 75
7. 15
8. 36
9. 15
10. 54
11. 100
12. 11
13. 8
14. -39
15. 400

PERENNIAL MATH

Grade 7-8 Individual Event (30 Minutes)

1) The first stage of a rocket burns 32 seconds longer than the second stage. If the total burning time for both stages is 148 seconds, how many seconds does the first stage burn?

2) In the figure shown the circle has a radius of 10 cm. What is the area of the square who's corners are all points on the circle?

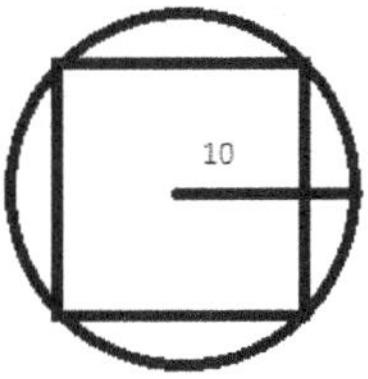

3) In the figure shown, all the shapes represent different numbers and the total sum of each row and column is shown. What is the sum of the top row?

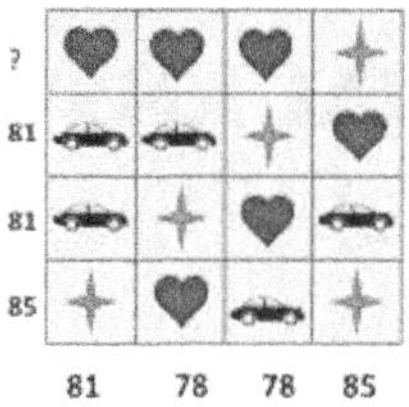

4) At his party, Tariq had 1087 jellybeans in his jar. For every 12 colorful jellybeans that he gave to friends, there was 1 licorice that he would throw away. He still had 34 jellybeans left in his jar after the party. How many jellybeans did he give away?

5) Bonnie's sales report for the month of February reflects that the ratio of roses to lilies sold was 6:1. If there were a total of 749 lilies sold, how many flowers did Bonnie sell in February?

6) The stadium has a parking lot that will hold 1000 vehicles. 2/5 of the parking spaces are for trucks. When Bob went to park his truck, there were 200 trucks and some cars in the parking lot. The parking lot was 3/4 full. How many cars were in it?

7) Akamu is twice as old as Kailani was three years ago. The total of their ages now is 36. How old is Kailani now?

8) Clyde counted the elves and reindeer at the Christmas photo booth. He noticed there were three and a half times the number of legs as the number of heads. The total number of legs and heads was 72. How many reindeer were counted?

9) A class of 50 students completed a survey on what pets they like. Everyone liked at least one pet. 4 students liked cats and fish but not dogs. 12 students liked dogs and fish but not cats. 10 students liked cats and fish. 18 students liked cats and dogs. Altogether, 30 students liked cats, 32 students liked dogs, and 28 students liked fish. How many students like fish only?

10) Joe can mow his mother's lawn in 60 minutes. His brother Ricky can mow it in 90 minutes. How many minutes will it take them to do it together?

PERENNIAL MATH

11) 1 + 2 + 3 +...+ 25 =

12) What is the ones digit of 7^{23}?

13) 24 – 3 x 4 + 12 ÷ 3 =

14) What is 25% of 15% of 280?

15) 9,999 x 999 =

INDIVIDUAL ANSWER DOCUMENT
(Grades 7-8)

SCORE

Team # __________ School ______________________________________

Student Name ____

Student Number ________________

1. ___________	6. ___________	11. ___________
2. ___________	7. ___________	12. ___________
3. ___________	8. ___________	13. ___________
4. ___________	9. ___________	14. ___________
5. ___________	10. ___________	15. ___________

PERENNIAL MATH

Grade 7-8 Individual Event Solutions

1) Let the second stage be x and the first stage be x + 32. Then (x + 32) + x = 148. 2x + 32 = 148. 2x + 32 – 32 = 148 – 32. 2x = 116. 2x/2 = 116/2. x = 58. The first stage is x + 32, so 58 + 32 = 90 seconds.

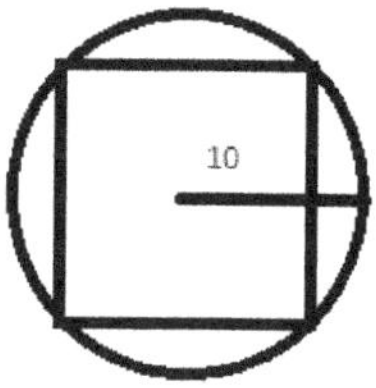

2) Since r = 10, one can make a 45^{0} 45^{0} 90^{0} right triangle which is a quarter of the square. The sides have a length of 10 cm because they are both a radius of the circle. Using the formula for area of a right triangle (Area=1/2 b x h) the area is ½ * 10 * 10 , or ½ * 100 = 50. Since this triangle is the area of ¼ the square, the total area is 4 * 50, or 200 cm^2.

3) Use algebra and let C=Car, S=Star, and H=Heart. Solve for the top row: 3H + S. First look for equations that have two of one item and one of the other two items and add them together.

2C + S + H = 81 (2nd row)
C + 2S + H = 85 (4th row)
C + S + 2H = 78 (3rd column)

4C + 4S + 4H = 244 divide both sides by 4 to get C + S + H = 61.

So, now subtract this equation from the original equations to solve for H and S.

C + S + 2H = 78	C + 2S + H = 85
- C + S + H = 61	- C + S + H = 61
H = 17	S = 24

Top row equals 3H + S or 3(17) + 24 = 75.

4) Take the total and subtract how many were left over. Divide this by thirteen because he gave away 12 and threw 1 away which is 13 portions.

N = (1087 - 34) ÷ 13
N = 81

Multiply 81 by 12 because he gave away 12 times as many as he threw away. He gave away 972 jellybeans.

5) Since the ratio of roses to lilies is 6:1, then there were 749 x 6 = 4494 roses. The total amount of flowers for February is 4494 roses + 749 lilies = 5243 flowers.

6) Draw a picture. Draw the parking lot in fifths and then separate these into fourths. The parking lot is ¾ full which is 15/20. Each fourth holds 50 since 50 x 20 = 1000. There would be 15 lots filled and 5 lots empty. There are 11 lots with cars. 11 x 50 = 550 cars.

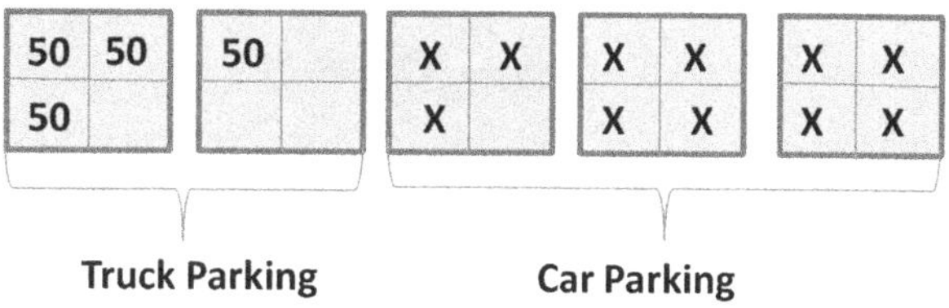

Grade 7-8 Individual Event Solutions

7) Akamu is twice as old as Kailani was three years ago. The total of their ages is 36. Akamu = 2 (K - 3) so solve for K:

$$2(K - 3) + K = 36$$
$$2K - 6 + K = 36$$
$$3K - 6 = 36$$
$$3K - 6 + 6 = 36 + 6$$
$$3K = 42$$
$$K = 42/3 \text{ or } 14$$

Akamu	Kailani	Total (36)
2 (K – 3)	K	36
22	14	36

So K - 3 = 11. Kailani is 14 now and was 11 years old 3 years ago. Akamu is 2(K-3) or 2(11) = 22 years old.
22 + 14 = 36 years old. Kailani is 14.

8) Use algebra or make a chart. There are 3 ½ times as many legs as heads. Let X represent heads. 3.5x + x = 72, 4.5x = 72 x= 72/4.5 = 16. So there are 16 heads. Working from 16 heads, you can use your chart to calculate legs. There are 12 Reindeer and 4 elves.

Elves	4
Reindeer	12
Heads	16
Legs	56
Total Heads + Legs (72)	72

9) All regions total 50 students. Label the totals for each category: 30 students liked cats, 32 liked dogs, and 28 liked fish. Fill in these areas: 4 students liked cats and fish but not dogs; 12 students liked dogs and fish but not cats. Use these to solve for missing areas. 10 students liked cats and fish [Region V + 4]. Region V = 6. 18 students liked cats and dogs [Region IV + Region V]. Region IV = 18 - 6 which is 12. Finally, solve for Region II to get the number of students who like fish only. 28 - 4 - 6 - 12 = 6.

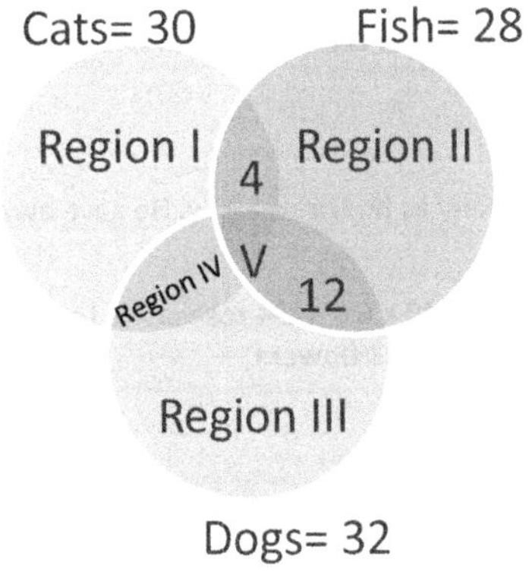

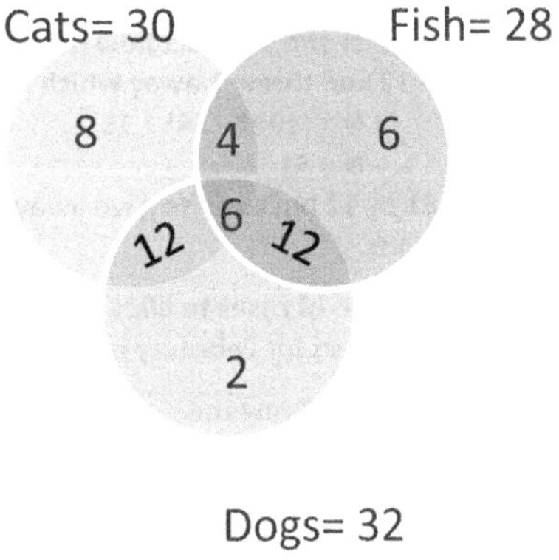

10) $$\text{Time} = \frac{1}{\frac{1}{X} + \frac{1}{Y}} = \frac{1}{\frac{1}{60} + \frac{1}{90}} = \frac{1}{5/180} = 36 \text{ minutes}$$

PERENNIAL MATH

Grade 7-8 Individual Event Solutions

11) 1 + 2 + 3 +...+ 25 =
Use the following formula Sum of 1 to n = n(n+1)/2
= 25(25 + 1)/2
= (25 x 26)/2
= 650/2
= 325

12) What is the ones digit of 7^{23} ? There is a pattern that repeats every fourth power:
$7^1 = \underline{7}$
$7^2 = 4\underline{9}$
$7^3 = 34\underline{3}$
$7^4 = 240\underline{1}$
$7^5 = 1680\underline{7}$
$7^6 = 11764\underline{9}$
...
...
To find the ones digit of 7^{23} divide 23rd power by 4 and you get a remainder of 3. So the ones digit would be 3 which is the third digit in the pattern 7, 9, $\underline{3}$, 1.

13) 24 – 3 x 4 + 12 ÷ 3 =
Use the order of operations PEMDAS.
24 – (3 x 4) + (12 ÷ 3) =
24 – 12 + 4 = 16

14) What is 25% of 15% of 280?
280 x .15 = 42
42 x .25 = 10.5

15)
```
 9,999
 x 999
 89991
 89991
89991
9,989,001
```

INDIVIDUAL ANSWER KEY

(Grades 7-8)

Instructions for Scoring:

- *Mark CORRECT answers by circling "O" the question number, and INCORRECT answers with an "X" . Each correct answer is worth one point. Count the circles (correct answers) and write that number in the "SCORE" box.*

CORRECT → (1.) ________

INCORRECT → X2. ________

1. 90
2. 200
3. 75
4. 972
5. 5,243
6. 550
7. 14
8. 12
9. 6
10. 36
11. 325
12. 3
13. 16
14. 10.5
15. 98,901

PERENNIAL MATH

TEAM TESTS GRADES 7-8

20 minute practice tests for teams
Each member gets a copy of the test, but the team only receives one answer document

Grade 7-8 Team Event (20 Minutes)

1) What is the sum of all the odd prime numbers 1 through 50?

2) Use the scale to solve the problem. How much does 1 apple, 1 orange, and 1 strawberry weigh altogether?

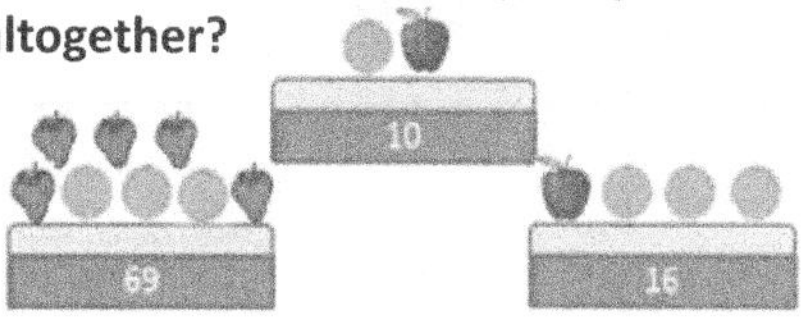

3) How many rectangles of any size are in this diagram?

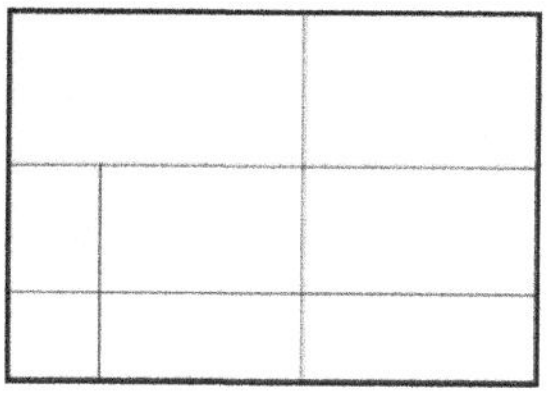

4) If two mosquitos can lay 8 eggs in 3 hours, how many eggs can 5 mosquitos lay in 6 hours?

5) Amy, Beverly and Carrie go shopping. When they pull out their money they discover that Amy and Beverly have $40 together, Beverly and Carrie have $57 combined, and Amy and Carrie have $43 together. How much money does Beverly have?

6) There are 5 purple, 4 pink, 3 blue, and 2 yellow gumballs in the machine. If Christina wants to be sure she gets a gumball of each color, how much money will she have to be willing to pay if each gumball costs 25 cents?

7) Solve the magic square. All rows, columns, and diagonals will have the same sum. What number will N be?

N		15	1
	7	6	
	11	10	8
	2	3	

8) Crazy Crème sells sundaes with the choice of fudge, caramel and strawberry toppings. One day they sold 263 sundaes; 144 had caramel, 144 had fudge, 86 had fudge and strawberry, 64 had caramel but no strawberry, 70 had fudge and caramel, 39 had all three toppings and 20 had no toppings. How many sundaes were sold with only strawberry on them?

9) The sum of three numbers A, B, and C is 96. The ratio of A:B is 5:4 and the ratio of B:C is 4:3. What is the difference between A and C?

10) There are 25 people at a meeting. If each person shakes hands only once with everyone else at the meeting, how many total handshakes will take place?

PERENNIAL MATH

TEAM ANSWER DOCUMENT
(Grades 7-8)

SCORE

Team # __________ School ______________________________

Student Names ____

____

Write answers clearly. Each correct answer is worth one point.

1. ____________

2. ____________

3. ____________

4. ____________

5. ____________

6. ____________

7. ____________

8. ____________

9. ____________

10. ____________

PERENNIAL MATH

Grade 7-8 Team Event Solutions

1) The sum of all the odd prime numbers 1-50 is 3 + 5 + 7 + 11 +13 + 17 + 19 + 23 + 29 + 31 + 37 + 41 + 43 + 47 = 326

2) Use algebra.

$$\begin{array}{l} (3O + A) = 16 \\ \underline{-\ (O + A) = 10} \\ 2O \quad = 6,\ \ O = 3 \end{array}$$

(O + A) = 10, A = 7
3O + 5S = 69, 3(3) + 5S = 69, 5S = 60, S = 12
O + A + S = 3 + 7 + 12 = 22

3) How many rectangles of any size are in this diagram?

Type	How Many
Made of 1 shape	8
Made of 2 shapes	9
Made of 3 shapes	4
Made of 4 shapes	2
Made of 5 shapes	2
Made of 6 shapes	1
Made of 8 shapes	1
Total rectangles	**27**

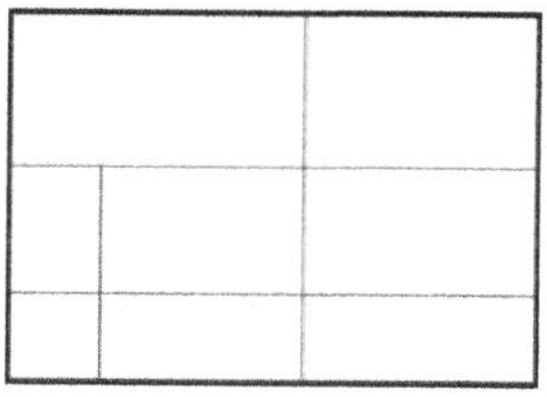

4) If two mosquitos can lay 8 eggs in 3 hours, then 1 mosquito can lay 4 eggs in 3 hours. Five mosquitos lay 20 eggs in 3 hours and 40 eggs in 6 hours.

5) Use Algebra and let A = Amy, B = Beverly and C = Carrie.

$$\begin{array}{l} A + B \quad = 40 \\ A + \quad C = 43 \\ \underline{\quad B + C = 57} \\ 2A+2B+2C=140 \end{array}$$

Now divide in half to get A + B + C = 70
Substitute the original equations back in (40) + C = 70, C = 30.
Since B + C = 57, then B + 30 = 57, B = 27.

6) Since there are 5 purple, 4 pink, 3 blue, and 2 yellow gumballs, she may have to get up to 13 gumballs before she gets 1 of each color because they could come out of the machine in order 5 purple, 4 pink, 3 blue, and then yellow. So she must be willing to pay 13 x .25 = $3.25

PERENNIAL MATH

Grade 7-8 Team Event Solutions

7) Column 3 is completed and all rows, and columns, and diagonals add up to 34. Select a square that is shared and use algebra. First solve for the bottom right square:

N		15	1
	7	6	
	11	10	8
X	2	3	13

X + 11 + 6 + 1 = X + 2 + 3 + ?
X + 18 = X + 5 + 13
Now solve for N:
34 - (13 + 10 + 7) = 4

4	14	15	1
9	7	6	12
5	11	10	8
16	2	3	13

8) Use a Venn Diagram. All regions and 1 outside region must add up to 263. The middle region that is intersected by all 3 rings is 39. Use this region to solve for the other regions.
The 2 regions that are shared by fudge and strawberry must add up to 86. 86 – 39 = 47.
The two regions shared by fudge and caramel add up to add up to 70. 70 - 39 = 31.
The two regions in caramel that are not shared with strawberry add up to 64. 64 – 31 = 33
The four regions for caramel must add up to 144.
144 – 31 – 39 – 64 = 41.
The four regions for fudge must add up to 144.
144 – 31 – 39 – 47 = 27.
Now find strawberry only by subtracting all other regions from total: 263 – (27 + 47 + 31 + 39 + 33 + 41 + 20) = 25
To check add all regions:
27 + 47 + 31 + 39 + 33 + 41 + 25 + 20 = 263 sundaes.

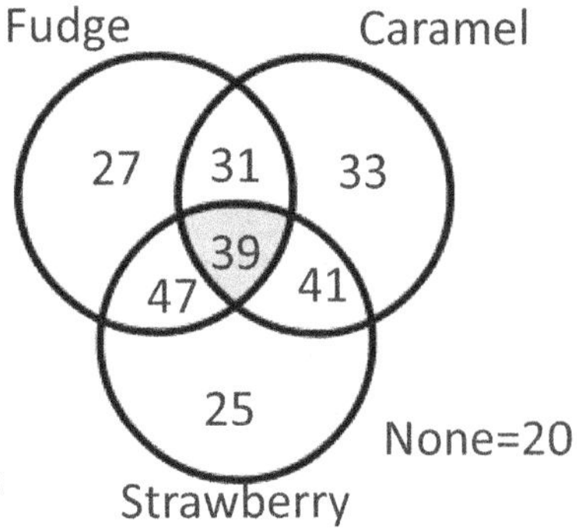

9) The sum of three numbers A, B, and C is 96. The ratio of A:B is 5:4 and the ratio of B:C is 4:3. There are 12 portions in A + B + C. 96/12 = 8. Each portion is worth 8. The difference between A and C is 5:3 or 40 – 24 = 16.

10) The first student (A) can shake hands with 24 other students. The second student (B) can shake hands with only 23 other students. They have already shaken hands with student A so that leaves 23 other students with whom they can shake hands. Using the same argument, the third student (C) can shake hands with 22 students (A and B have already been greeted and student C does not shake hands with himself). The total number of handshakes will be 24 + 23 + 22 + 21 + ...+3 + 2 + 1 = 300.

TEAM ANSWER KEY
(Grades 7-8)

Instructions for Scoring:

- *Mark CORRECT answers by circling "O" the question number, and INCORRECT answers with an "X" . Each correct answer is worth one point. Count the circles (correct answers) and write that number in the "SCORE" box.*

CORRECT → (1.) ________

INCORRECT → X. ________

1. **326**
2. **22**
3. **27**
4. **40**
5. **27**
6. **3.25**
7. **4**
8. **25**
9. **16**
10. **300**

PERENNIAL MATH

Grade 7-8 Team Event (20 Minutes)

1) What is the product of the sum of the first 7 even numbers and the sum of the first 5 prime numbers greater than 20?

2) The area of each of the 8 rectangles in this figure is $18cm^2$ and their lengths are twice the width. What is the perimeter of the entire figure that is made up of these smaller rectangles?

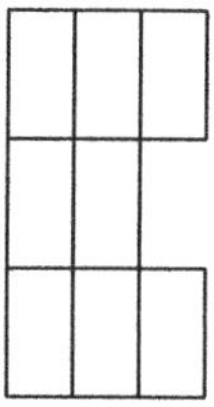

3) In this magic square the rows, columns, and diagonals all have the same sum. What number goes in the center square?

17		13
7	?	15
	19	

4) The video game club purchased a new game system. Each of the 6 members paid the same amount. When 2 new members joined and put in money on the game system, each of the original 6 members received a check back for $10. How much did the game system cost?

5) If the day before the day before yesterday was Monday, what day of the week will four days after tomorrow be?

6) Pizza Heaven sells pizzas with the choice of pepperoni, sausage, and ham for toppings. One day they sold 128 total pizzas. Of these, 80 had pepperoni, 80 had sausage, 30 had pepperoni and ham only, 40 had sausage but no ham, 53 had only pepperoni and sausage only, 28 had all three toppings and 15 had no toppings. How many pizzas were sold with only ham on them?

7) Arthur bought a coin for $250, sold it for $320, bought it back for $400 and finally sold it for $380. How many dollars did Arthur make or lose? (You must write "made" or "loss" in your answer)

8) There were 10 people at the party. Each person shook hands with everyone else only once. How many total handshakes were exchanged?

9) If the average of three positive integers is 6, the greatest possible value of the sum of the squares of the three integers is ______?

10) Trent can buy 2 pairs of pants and 1 shirt for $79.85 and 1 pair of pants and 2 shirts for $69.85, how much does one shirt cost?

 PERENNIAL MATH

TEAM ANSWER DOCUMENT
(Grades 7-8)

SCORE

Team # __________ School ____________________________________

Student Names____

____

Write answers clearly. Each correct answer is worth one point.

1. ____________

2. ____________

3. ____________

4. ____________

5. ____________

6. ____________

7. ____________

8. ____________

9. ____________

10. ____________

PERENNIAL MATH

Grade 7-8 Team Event Solutions

1) The sum of the first 7 even numbers is (2 + 4 + 6 + 8 + 10 + 12 + 14) = 55. The sum of the first 5 prime numbers > 20 is (23 + 29 + 31 + 37 + 41) = 161. Their product is 56 x 161 = 9,016.

2) The area of each of the 8 rectangles in this figure is 18cm^2 and their lengths are twice the width. So the length is 6cm and the width is 3cm. The perimeter of the entire figure made up of these smaller rectangles is equal to the number of 3cm sides + the number of 6cm sides. (8 x 3cm = 24cm) + (6 x 6cm = 36cm). The perimeter is 24cm + 36cm= 60cm.

3) In this magic square the rows, columns, and diagonals all have the same sum. Select a shared square (X) and solve for the adjacent numbers. For instance, the middle square is shared by every row, column and diagonal. So 7 + ? + 15 = 13 + ? + X. X = 9. Now the sum of each row column and diagonal is 17 + 7 + 9 = 33. The center box is equal to 7 + ? + 15 = 33 or 33 – 22 = 11.

17		13
7	?	15
X	19	

4) The 6 original members paid the same amount and received a check back for $10 when 2 more joined. So the amount given back was $60. The 2 new members put in $60, so divide that by 2 to see how much each person paid. Each person paid $30. The total cost of the game system was $30 x 8 members = $240.

5) If the day before the day before yesterday was Monday, what day of the week will four days after tomorrow be? Make a timeline. It will be Tuesday.

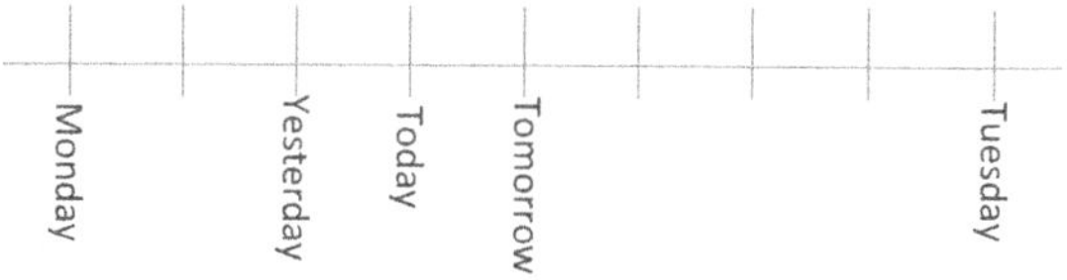

6) Use a Venn Diagram. The 15 goes on the outside because it represents the pizza with no toppings. Both pepperoni and sausage have 80. 28 had all three toppings goes in the center. 30 had pepperoni and ham only, so region IV should be 30 – 28 = 2. 53 had pepperoni and sausage only, so region II should be 53 – 28 = 25. The P circle should total up to 80. So region I is 80 – 25 – 28 – 2 = 25. 40 had sausage but no ham which is region II + III. So 40 – 25 = 15 for region III. 80 had sausage so region V is 80 – 25 – 15 – 28 = 12. There are 128 total pizzas. So, subtract all of the regions to solve for region VI. 128 – 15 – 25 – 25 – 15 – 2 – 28 – 12 = 6. There were 6 pizzas with only ham (region VI).

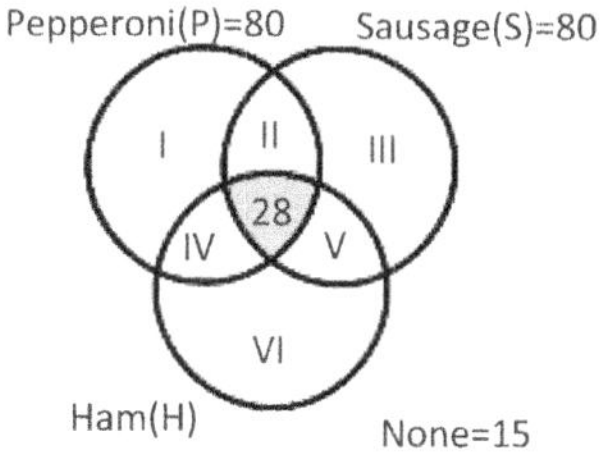

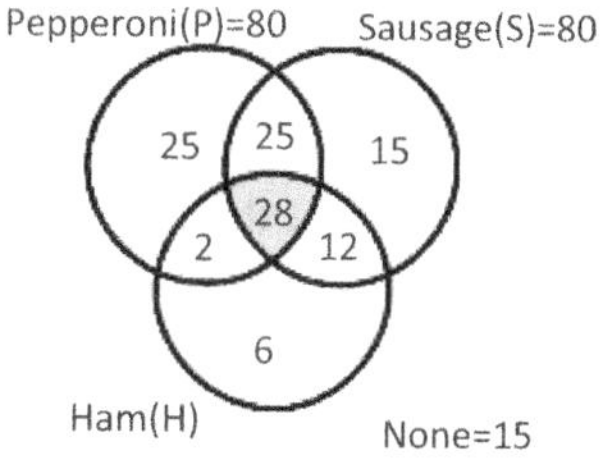

Grade 7-8 Team Event Solutions

7) Act it out. Suppose Arthur starts with \$500 and bought a coin for \$250 he would have \$250 left. Now he sold the coin for \$320 and he would have \$570 in his wallet. He buys that coin back for \$400 and he has \$170 left. Finally he sold it for \$380 and he has \$550 in his wallet. Since he started with \$500, he made a profit of \$50. So, write "made 50" as your answer.

8) There were 10 people at the party. Each person shook hands with everyone else only once. The first person would shake hands with 9 other people. The second with 8 and so forth. 9 + 8 + 7 + 6 + 5 + 4 + 3 + 2 + 1 = 45 handshakes.

Person 1	Person 2	Person 3	Person 4	Person 5	Person 6	Person 7	Person 8	Person 9
Person 2	Person 3	Person 4	Person 5	Person 6	Person 7	Person 8	Person 9	Person 10
Person 3	Person 4	Person 5	Person 6	Person 7	Person 8	Person 9	Person 10	
Person 4	Person 5	Person 6	Person 7	Person 8	Person 9	Person 10		
Person 5	Person 6	Person 7	Person 8	Person 9	Person 10			
Person 6	Person 7	Person 8	Person 9	Person 10				
Person 7	Person 8	Person 9	Person 10					
Person 8	Person 9	Person 10						
Person 9	Person 10							
Person 10								

9) If the average of these integers is 6, then their sum is 18, and the greatest possible value of the sum of their squares is $1^2 + 1^2 + 16^2 = 258$.

10) Use algebra. Let p = pants and s = shirts.

$$\begin{array}{rl} & 2p + s = 79.85 \\ + & \underline{p + 2s = 69.85} \\ & 3p + 3s = 149.70 \end{array}$$

Divide both sides by 3.

$$p + s = 49.90$$

Subtract out a pair of pants and shirt from 1 pair of pants and a shirt to solve for the cost of 1 shirt.

$$\begin{array}{rl} & p + 2s = 69.85 \\ - & \underline{p + s = 49.90} \\ & s = 19.95 \end{array}$$

TEAM ANSWER KEY
(Grades 7-8)

Instructions for Scoring:

- *Mark CORRECT answers by circling "O" the question number, and INCORRECT answers with an "X" . Each correct answer is worth one point. Count the circles (correct answers) and write that number in the "SCORE" box.*

CORRECT → (1.) ________

INCORRECT → X. ________

1. 9,016
2. 60
3. 11
4. 240
5. Tuesday
6. 24
7. made 50
8. 45
9. 258
10. 19.95

PERENNIAL MATH

Grade 7-8 Team Event (20 Minutes)

1) In the dress shop the ratio of blue dresses to the number of red dresses is 7:13. What percent of the dresses are blue?

2) How many degrees are in Angle A?

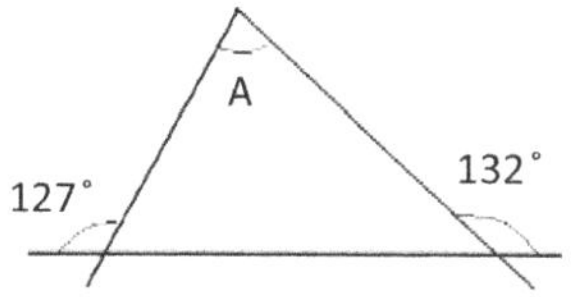

3) Find the area of the shaded region.

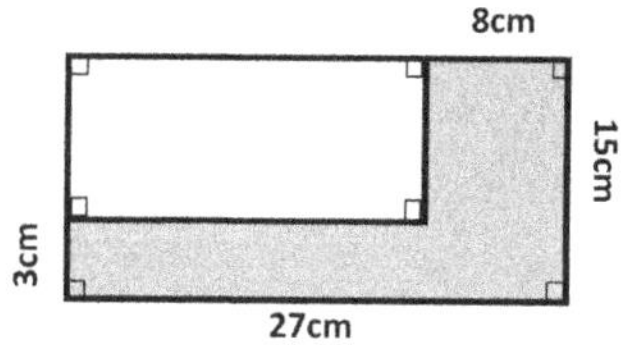

4) Meredith is twice as old as Jackson was three years ago. The total of their ages now is 36. How old is Jackson now?

5) What is the sum of all the prime numbers less than 100 that contain the digit 7?

6) Trent can buy 2 pencils and 1 eraser for 97 cents and 1 pencil and 2 erasers for 83 cents. How much does one eraser cost?

7) In a tennis tournament there are 128 players in the draw. If this is a single elimination tournament, how many matches will the champion have to play in order to hold the championship trophy?

8) Clyde counted the flamingos and giraffes at the zoo. He noticed there were three and a half times the number of legs as the number of heads and the total of the legs and heads was 72. How many giraffes were counted?

9) A class of 50 students completed a survey on what music they like. Everyone liked at least one type. Four students liked rap and pop, but not rock. Twelve students liked rock and pop, but not rap. Ten students liked rap and pop. Eighteen students liked rap and rock. Altogether, 30 students liked rap, 32 students liked rock, and 28 students liked pop. How many students liked pop only?

10) Using any combination of half dollars, quarters, dimes, and nickels how many ways can you make 85 cents?

TEAM ANSWER DOCUMENT
(Grades 7-8)

SCORE

Team # __________ School ____

Student Names ____

____

Write answers clearly. Each correct answer is worth one point.

1. ____________

2. ____________

3. ____________

4. ____________

5. ____________

6. ____________

7. ____________

8. ____________

9. ____________

10. ____________

PERENNIAL MATH

Grade 7-8 Team Event Solutions

1) The ratio of blue dresses to the number of red dresses is 7:13. Assume there are 20 total dresses, so 7 are blue and 13 are red. Therefore, the percent that of blue dresses is 7/20 or 35/100 which is 35%.

2) Subtract to find the values of the inside angles first. $180^\circ - 127^\circ = 53^\circ$ and $180^\circ - 132^\circ = 48^\circ$. Now add those angles and subtract from 180°. $180^\circ - (53^\circ + 48^\circ) = 79^\circ$

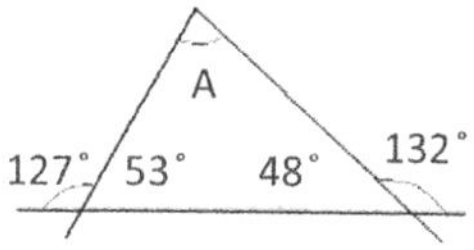

3) Break the shape up into smaller shapes and find the areas of the smaller shapes and add them up. One of the areas is 8cm x 15cm = $120cm^2$. The next area is (27 - 8 = 19cm) x 3cm. 19cm x 3cm = $57cm^2$. Now add those smaller areas together: $120cm^2$ + $57cm^2$ = $177cm^2$

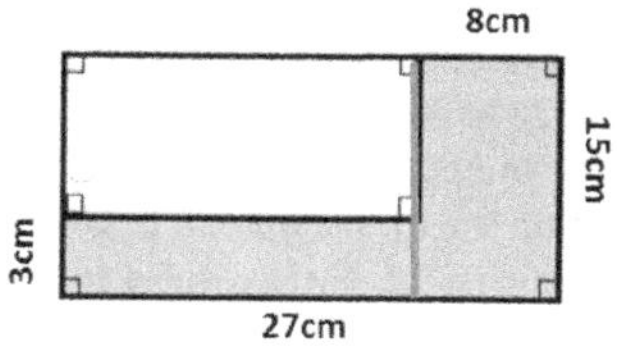

4) Use algebra. Express Meredith in terms of Jackson, so Meredith = 2 (J -3).

$2(J-3) + J = 36$
$2J - 6 + J = 36$
$3J - 6 = 36$
$3J - 6 + 6 = 36 + 6$
$3J = 42$
$J = 42/3 = 14$

Jackson is 14 now. Check answer by calculating Meredith's age now. Remember that Meredith is 2(J-3) or 2(11) = 22 years old, which is twice as old as Jackson's age 3 years ago.

5) The sum of all the primes under 100 with a digit 7 in them is:

$$7 + 17 + 37 + 47 + 67 + 71 + 73 + 79 + 97 = 495$$

6) Use algebra.

$2p + 1e = .97$
$+\ \underline{1p + 2e = .83}$
$3p + 3e = 1.80$ Divide both sides by 3
$p + e = .60$

Subtract this from one of the known equations above to determine the value of the eraser

$1p + 2e = .83$
$-\ \underline{p + e = .60}$
$e = .23$

7) Start with 128 players and divide by 2 for each match played until you get to one, the champion:

128 ÷ 2 = 64; 64 ÷ 2 = 32; 32 ÷ 2 = 16; 16 ÷ 2 = 8; 8 ÷ 2 = 4; 4 ÷ 2 = 2; 2 ÷ 2 = 1.

There are 7 games that the winner must play.

PERENNIAL MATH

Grade 7-8 Team Event Solutions

8) Use algebra. There are 3.5 times as many legs as heads. Let x represent heads. So the total legs and heads can be represented with the equation: $3.5x + x = 72$, $4.5x = 72$, $x = 72/4.5 = 16$ heads. Working from 16 heads, you can make a chart to calculate legs. Start with 8 of each animal and notice that the total heads and legs is less than 72, so increase the number of giraffes (decrease the number of flamingos) each time until you reach 72. There are 12 giraffes and 4 flamingos.

Flamigos	8	6	4
Giraffes	8	10	12
Heads (H)	16	16	16
Legs (L)	48	52	56
Total (H+L)	64	68	72

9) All regions total 50 students. Label the totals for each category: 30 students liked rap, 32 liked rock, and 28 liked pop. Fill in these areas: 4 students liked rap and pop but not rock; 12 students liked rock and pop but not rap. Use these to solve for missing areas. 10 students liked rap and pop [Region V + 4]. Region V = 6. 18 students liked rap and rock [Region IV + Region V]. Region IV = 18 - 6 which is 12. Finally, solve for Region II to get the number of students who like pop only. 28 - 4 - 6 - 12 = 6.

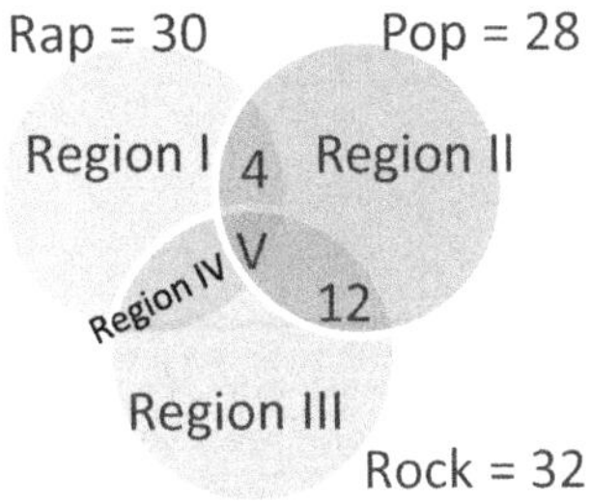

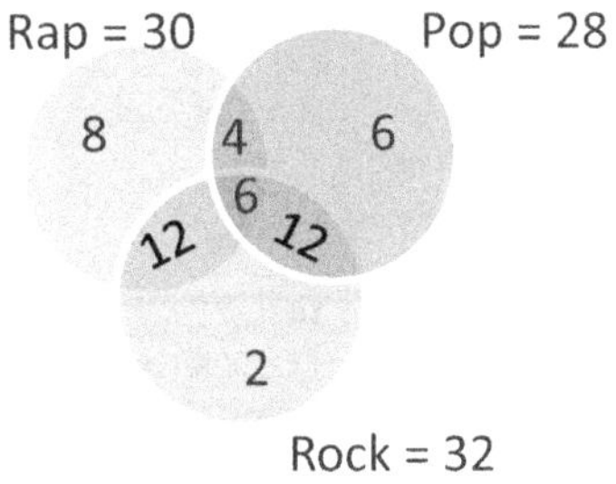

10) Make an organized list. There are 28 combinations.

Half Dollar	Quarters	Dimes	Nickels
1	1	1	0
1	1	0	2
1	0	3	1
1	0	2	3
1	0	1	5
1	0	0	7
0	3	1	0
0	3	0	2
0	2	3	1
0	2	2	3
0	2	1	5
0	2	0	7
0	1	6	0
0	1	5	2
0	1	4	4
0	1	3	6
0	1	2	8
0	1	1	10
0	1	0	12
0	0	8	1
0	0	7	3
0	0	6	5
0	0	5	7
0	0	4	9
0	0	3	11
0	0	2	13
0	0	1	15
0	0	0	17

TEAM ANSWER KEY
(Grades 7-8)

Instructions for Scoring:

- ***Mark CORRECT answers by circling "O" the question number, and INCORRECT answers with an "X" . Each correct answer is worth one point. Count the circles (correct answers) and write that number in the "SCORE" box.***

CORRECT → (1.) ________

INCORRECT → X. ________

1. **35**
2. **79**
3. **177**
4. **14**
5. **495**
6. **.23**
7. **7**
8. **12**
9. **6**
10. **28**

PERENNIAL MATH

Grade 7-8 Team Event (20 Minutes)

1) What is the sum of all the odd composite numbers 1 through 35?

2) Use the scale to solve the problem. How much does 1 apple, 1 orange, and 1 pear weigh altogether?

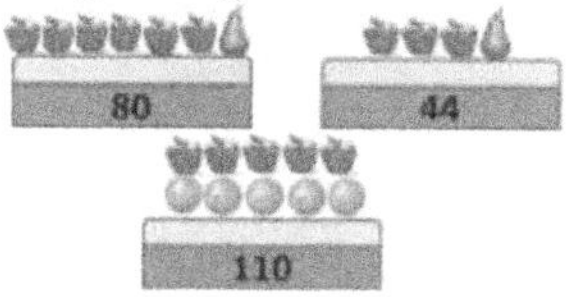

3) How many rectangles of any size are in this diagram?

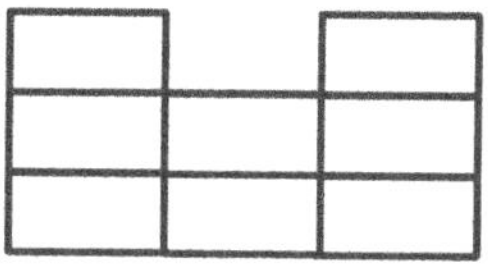

4) How many different coin combinations will give you 95 cents, if there are no pennies and more dimes than nickels? Quarters and half dollars may also be used.

5) Meredith bought 4 pairs of pants, 3 shirts, and 2 pairs of shoes. How many different outfits can she make?

6) How many positive factors of 400 are divisible by 4?

7) Three small congruent circles, surrounded by a large circle, have their centers on a diameter of the large circle. If the area of the shaded region is 96π, what is the area of one of the small circles? Express your answer in terms of π.

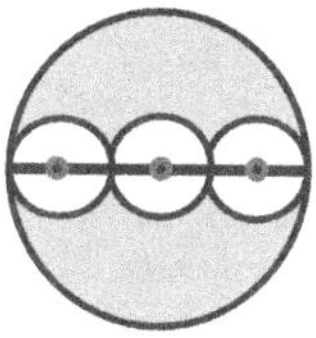

8) Super Dog sells hotdogs with the choice of ketchup, mustard, and relish. One day they sold 256 hotdogs. 140 had mustard, 140 had ketchup, 84 had ketchup and relish only, 62 had mustard but no relish, 68 had ketchup and mustard only, 38 had all three condiments, and 20 had no condiments. How many hotdogs were sold with only relish on them?

9) In the basketball league the ratio of teams that wear blue uniforms to the number of teams that wear red uniforms is 12:13. What percent of the teams in the league wear blue uniforms?

10) There are 20 people at a meeting. If each person shakes hands only once with everyone else at the meeting, how many total handshakes will take place?

Grade 7-8 Team Event Solutions

1) The sum of all the odd composite numbers 1-35 is 9 + 15 + 21 + 25 + 27 + 33 + 35 = 165.

2) (3A + P) + 3A = 80
 <u>- (3A + P)   = 44</u>
        3A = 36, or A = 12

Next:
 5A + 5O = 110 (Substitute A = 12)
5(12) + 5O = 110,
 110 - 60 = 50, so O = 10
Substitute 12 in for A to solve for P
If 3A + P = 44, then 3(12) + P = 44, 44-36 = 8 So P = 8
O + A + P= 10 + 12 + 8 = 30

(3A + P) + 3A = 80   (3A + P) = 44

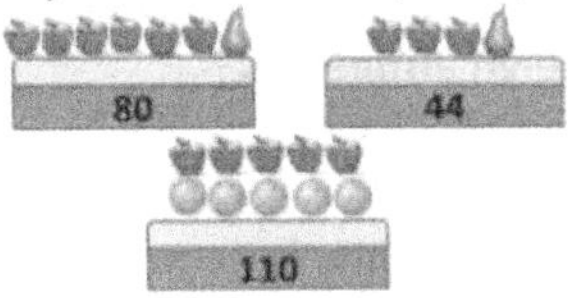

3) How many rectangles of any size are in this diagram?

Type	How Many
Made of 1 shape	8
Made of 2 shapes	9
Made of 3 shapes	4
Made of 4 shapes	2
Made of 6 shapes	1
Total Rectangles	24

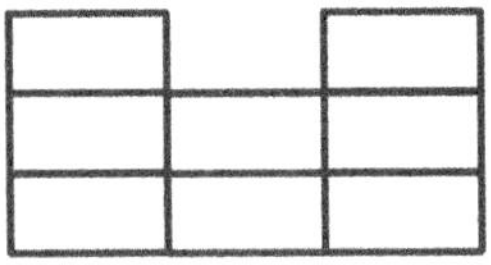

4) Make a chart to see the different coin combinations for 95 cents that use half dollars, quarters, dimes and nickels and more dimes than nickels. There are 9 combinations.

Half Dollar	Quarter	Dime	Nickel	Total
1	1	2	0	0.95
1	0	4	1	0.95
0	3	2	0	0.95
0	2	4	1	0.95
0	1	7	0	0.95
0	1	6	2	0.95
0	0	9	1	0.95
0	0	8	3	0.95
0	0	7	5	0.95

5) Meredith bought 4 pairs of pants, 3 shirts and 2 pairs of shoes. How many different outfits can she make?

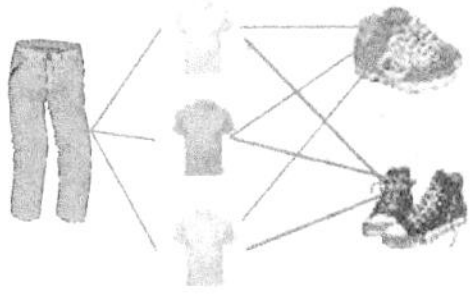

6 outfits for each pair of pants x 4 pairs of pants = 24 outfits.

Use multiplication principle:
4 x 3 x 2 = 24

Grade 7-8 Team Event Solutions

6) Make a list of factors of 400 and see which factors are divisible by 4. There are 9.

1	400
2	200
4	100
5	80
8	50
10	40
16	25
20	20

7) We could use guess and check until the area of the large circle = 96π.
-Guess #1: If the small circle has r = 1, then the large circle has R = 3, and $\pi R^2 - 3\pi r^2 = 9\pi - 3\pi = 6\pi$.
-Guess #2: If the small circle has r = 2, then the large circle has R = 6, and $\pi R^2 - 3\pi r^2 = 36\pi - 12\pi = 24\pi$.
-Guess #3: If the small circle has r = 3, then the large circle has R = 9, and $\pi R^2 - 3\pi r^2 = 81\pi - 9\pi = 72\pi$.
-Guess #4: If the small circle has r = 4, then the large circle has R = 12, and $\pi R^2 - 3\pi r^2 = 144\pi - 48\pi = 96\pi$. So, the area of one small circle with r = 4 is $\pi r^2 = 4^2\pi = 16\pi$.

8) Use a Venn Diagram. The 20 goes on the outside because it represents the burgers with nothing. Both ketchup and mustard have 140. 38 had all three condiments goes in the center. 84 had ketchup and relish. The total number in region IV should be 84, so 84 – 38 = 46. 68 had ketchup and mustard only, so region II should be 68 – 38 = 30.
The K circle should total up to 140. So region I is 140 – 30 – 38 – 46 = 26. 62 had mustard but no relish which is region II + III. So 62 – 30 = 32 for region III. 140 had mustard so region V is 140 – 32 – 38 – 30 = 40. There are 256 total hotdogs. So, subtract all of the regions to solve for region VI. 256 – 20 – 40 – 46 – 38 – 30 – 32 – 26 = 24. There were 24 hotdogs with only relish (region VI).

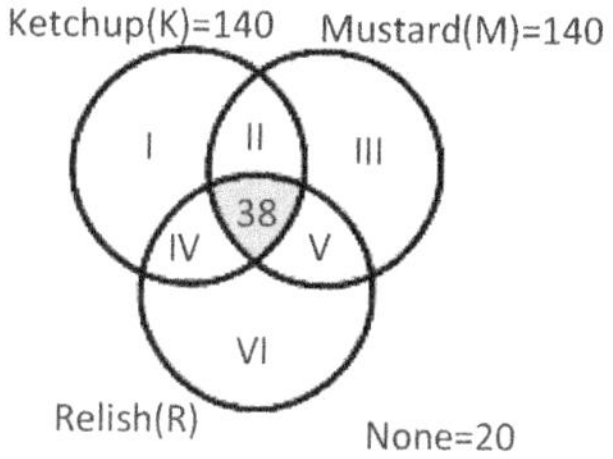

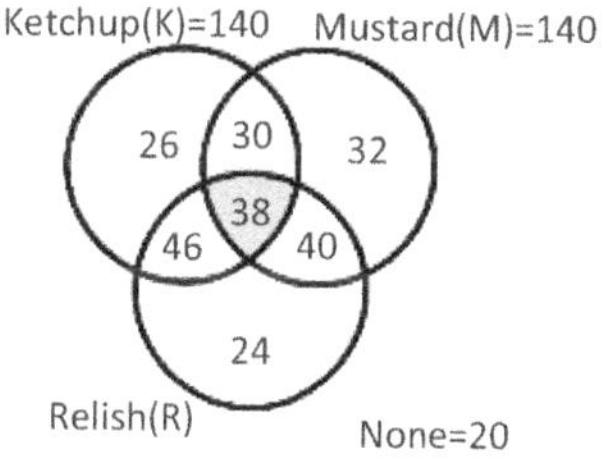

9) The ratio of teams that wear blue uniforms to the number of teams who wear red uniforms is 12:13. Of the 25 total teams, 12 wear blue uniforms. So the percent that wear blue uniforms is 12/25 or 48/100 which is 48%.

10) The first person (A) can shake hands with 19 other people. The second person (B) can shake hands with only 18 other people. They have already shaken hands with person A so that leaves 18 other people with whom they can shake hands. Using the same argument, the third person (C) can shake hands with 17 people (A and B have already been greeted and person C does not shake hands with himself). The total number of handshakes will be
19 + 18 + 17 + 16 + ...+3 + 2 + 1 = 180.

TEAM ANSWER KEY
(Grades 7-8)

Instructions for Scoring:

- ***Mark CORRECT answers by circling "O" the question number, and INCORRECT answers with an "X" . Each correct answer is worth one point. Count the circles (correct answers) and write that number in the "SCORE" box.***

CORRECT → (1.) ________

INCORRECT → X. ________

1. **165**
2. **30**
3. **24**
4. **9**
5. **24**
6. **9**
7. **16π**
8. **24**
9. **48**
10. **180**

Made in the USA
Monee, IL
05 November 2026

67811041R00036